YOUR KNOWLEDGE HAS VALUE

- We will publish your bachelor's and
  master's thesis, essays and papers

- Your own eBook and book -
  sold worldwide in all relevant shops

- Earn money with each sale

Upload your text at www.GRIN.com
and publish for free

Evaluation of the Toxicological Effects of the Ethanolic Extract of Alstonia Boonei Stem Bark

Seun Akingbotolu

Bibliographic information published by the German National Library:

The German National Library lists this publication in the National Bibliography; detailed bibliographic data are available on the Internet at http://dnb.dnb.de.

ISBN: 9783346315632
This book is also available as an ebook.

Evaluation of the Toxicological Effects of the Ethanolic Extract of Alstonia Boonei Stem Bark

Seun Akingbotolu

Department of Biochemistry, Faculty of Science, Obafemi Awolowo University, Ile-Ife, Nigeria (BSc Biochemistry).

ABSTRACT

This study investigated the toxicological effects of ethanolic extract from alstonia boonei stem bark, based on the dosage and duration of administration to albino mice. We evaluated for the phytochemistry of the stem bark and the clinical biochemical parameters of blood plasma of albino mice. After administering the aqueous extract to the albino mice, the biochemical parameters in their blood plasma were determined using commercially available kits. The determination of the phytochemical constituent of the ethanolic extract revealed the presence of alkaloids, flavonoids, saponins, terpenoids and cardiac glycosides. After the first week of administration, no significant variation in most of the biomarkers was recorded. However, after second week of treatment, urea and alanine aminotransferase levels fell significantly. Consequently, the ratio of aspartate transaminase to alanine transaminase was found to be much greater than 2, which was attributed to alcoholic toxicity from the ethanolic extract consumption.

KEY WORDS: Anti-malarial, alstonia boonei, ethanolic extract, toxicology, albino mice

INTRODUCTION

Malaria is a life-threatening disease caused by a parasitic infection of the red blood cells and transmitted through a bite of the female anopheles mosquitoes (1). It is undoubtedly the single most destructive and dangerous infectious agent in the developing world, predominantly in tropical and subtropical regions, including parts of the Americans, Asia and Africa (1-4). The World Health Organisation and the World Bank rank malaria as the largest single component of the disease burden in Africa, causing an annual loss of 35 million future life-years from disability and premature mortality (5). In Africa, malaria is responsible for about 20-30% of hospital admissions and about 30-50% of outpatient consultation (3).

In recent times there has been an upsurge in the popularity and the use of plant extracts as an alternative form of treatment and prevention of malarial and various other related medical conditions (6). This could be attributed to the proven therapeutic efficacy, easy accessibility and affordability of herbal medications (7-9) or the growing rate of anti-microbial drug resistance to previously susceptible organisms (10-14). While these alternative forms of treatments have yielded results, their side effects when administered for a prolonged period have not been adequately addressed. As a result, there has been increasing interest in understanding the side effects of the prolonged use of these medications.

Astonia boonei, an abundant evergreen plant is one of the commonest medical plants with almost all its parts being used for the treatment of various diseases including malaria, intestinal helminths, hypertension, chronic diarrhoea and poisoning from snake bites (7, 8). In Nigeria and other west and central African countries where the consumption of this plant species as medicinal herb is very rampant, several pharmacological studies have been conducted to determine the efficacy of extracts from different parts of the plant including its leaves, flowers, stem root and bark (13, 14). These studies have revealed anti-malarial, anti-pyretic, analgesic, anti-inflammatory, diuretic, anti-helmintic and anti-hypertensive properties of the plant extract (13). The stem bark of alstonia boonei is currently listed among the anti-malarial drug in the African pharmacopeia.

A study carried out to evaluate the effect of ethanolic extract from alstonia boonei leaves on levels of serum electrolyte in wistar albino rats at dosages ranging from 100 to 400 mg/kg body weight showed no significant changes ($p<0.05$) in electrolyte balance at all doses administered (10). However, serum creatinine levels were significantly elevated ($p<0.05$) when doses higher than 400 mg/kg body weight was administered. Several other studies have also been conducted on various parts of alstonia boonei and its efficacy by administering low concentration doses of the extract to various animal specimen (13, 14). On the other hand, a study on its toxicology that reflects its effects for a prolonged use or high dosage of administration has not been adequately addressed. It is of great medical importance to investigate their effects when administered for a prolonged period because their cheap availability in the African nations makes them affordable and viable options for regular treatment of malaria and related diseases. In view of this, we report detailed study conducted to determine the toxicological effects of the prolonged use of ethanolic extract from alstonia boonei stem bark.

The remainder of this paper is organized into four distinct parts. First, the experimental section describes the animal groupings and plant extract preparations, tests methods and statistical analysis procedure used. Second, the results are presented, and third, the discussions are made by comparing this study with previous works. Lastly, our conclusions are briefly stated.

METHODS

Collection and preparation of plant extract and phytochemical screening

Fresh stem barks were collected from a large alstonia boonei tree at a location in Ile-Ife, Osun State Nigeria. The plant material was authenticated at the Herbanium in the faculty of pharmacy, Obafemi Awolowo University and deposited with the specimen number FPI-1886. Thereafter, the cut stem barks were air dried and then ground into fine powder at the drug research and production unit (DRPU), faculty of pharmacy, Obafemi Awolowo University Ile-Ife. About 500 g of the powder was weighed, diluted into two litres (2 L) of seventy percent (70%) ethanol and allowed to stand at room temperature for two days (48 hours). Afterward, the solution was sieved using a cheese cloth to obtain the supernatant which was further concentrated to dryness using a rotatory evaporator. This produced a brown residue that was termed the ethanolic extract (EE) of alstonia boonei. Using the EE, the phytochemical tests were carried out for the presence of phytochemicals responsible for the anti-microbial potential in the stem bark using standard procedure. Screening for the presence of the following chemicals were carried out: alkaloids, saponins, tannins, flavonoids, terpenoids, cardiac glycosides, phlobatannins and anthraquinone. Detailed description of the phytochemical test protocols has been made elsewhere (14)

Animal preparation, grouping and extract administration for toxicological study

Eighteen (18) albino mice having an average weight of 62-120 g were purchased from the animal house, College of Health Science, Obafemi Awolowo University Ille-ife. The animals were housed in a cage, had free access to rat food and water and were acclimatized for 14 days in the animal house under standard environmental condition. Thereafter, the mice were weighted and randomly grouped into three groups (A, B, C) with each group consisting of six mice. Mice in group A served as the control group and were given only distilled water. Mice in group B received 500 mg/kg body weight of the ethanolic extract, whereas mice in group C received 1000 mg/kg body weight of the extract. Ethanolic extract was administered orally to mice in groups B and C every day and after the first seven days, the first set of mice from groups A, B and C were sacrificed, while the remaining were fed for another seven days (now 2 weeks) and then the second set consisting of 2 mice from each group were sacrificed. The extract administration was then discontinued, and the remaining animals were left alone for another week (now 3 weeks in total) to observe for signs of recovery from previous extract examination.

In sacrificing the animals, all institutional guides for the care and use of laboratory animals were followed. The animals were first anaesthetized with diethyl ether, followed by direct cardiac puncture. The blood samples of the animals were then collected into heparinized sample bottles for chemical evaluation and analysis. Cut portions of the brain, spleen and liver of the mice were obtained and kept in 10% buffered formalin for histopathological evaluation which are not included in the current study.

Biochemical evaluation of blood samples

Blood samples were collected from the euthanized animals and centrifuged at 3000 rpm on 90-2 bench centrifuge for 10 minutes. The supernatant obtained with the randox diagnostic kit was stored in the refrigerator for further analysis. From the supernatant, albumin, aspartate aminotransferase (AST), alanine amino transferase (ALT), creatinine concentration, alkaline phosphate (ALP), bilirubin and urea concentration were estimated. Albumin was estimated using bromocresol green (BCG) modified method of Pinnell and Northan (15), AST and ALT were determined using the colorimetric method of Reitman and Frankel (16). Creatinine concentration was determined using Jaffe's-alkaline picrate method as described by Kume and co-workers (17). ALP was determined according to the method of Sanni and Van Etten (18). Bilirubin concentration was estimated using Gambino modified method of Jendrassik and Groff (19) and urea concentration was estimated using the colorimetric method of Berthelot as modified by Gordon and co-workers (20).

Statistical analysis

Data obtained from biochemical assay were analysed using one-way analysis of variance (ANOVA). The Turkey HSD multiple comparison test was used to verify differences between treatment groups. The result was expressed as mean± standard error of duplicates.

RESULTS

Phytochemical screening

The primary benefits plant-derived medicine offer includes affordability and better safety margin in comparison with synthetic alternatives. The findings in this study prove that the ethanolic extract (EE) from alstonia boonei stem bark contains important phytochemicals that support its use as a medicine with numerous health benefits. Moreover, the wide range of these phytochemicals makes the EE suitable for the treatment of several diseases such as malaria, fever, intestinal helminthes, rheumatism and hypertension. Our test results showed that only phlobatannins and anthraquinones were absent, which is consistent with that previously reported (14), whereas alkaloids, tannins, saponins, flavonoids cardiac glycoside (with or without steroidal ring) and terpenoids were all present. Although these phytochemicals essentially work in harmony with antioxidants such as vitamin C, E and selenium, they offer considerably greater protection against cancer, oxidative stress and tumour proliferation than the simple nutrients. Having observed the phytochemical constituents of the EE, we sought to analyse its toxicological effect when administered for a prolonged period by carrying out biochemical assay on the supernatant obtained from the albino mice.

Toxicological examination for urea and creatinine levels in the blood

Observations made during biochemical assay studies have proven that the intake of the EE of alstonia boonei bark at low (500 mg/kg body weight) and high doses (1000 mg/kg body weight) may have deleterious effect on one's overall health, especially when this concoction is consumed over a prolonged period of time. This finding is strengthened by results obtained from the quantitative analysis of circulating tissue enzymes in blood of the specimens.

Interestingly, table 1 shows that there were no significant changes in the blood urea level of the specimens when the EE was consumed for 7 days (21-23). In fact, the recorded changes in this parameter do not constitute a threat to one's life as they are within standard reference values for a healthy mouse as reported by Bray and Preston (22). Therefore, this extract does not alter urea levels when taken for a week, whether in low or high dose. On the other hand, the blood urea levels were observed to be significantly depleted when both low (500 mg/kg body weight) and high doses (1000 mg/kg body weight) were administered for two weeks as shown in table 2 (24, 25). However, following discontinuation of the extract administration for another week, the blood urea levels reverted to normal range (table 3). Although, urea levels in the blood are dependent on protein ingestion and renal function, this finding of reduced urea levels apparently suggests the occurrence of acute liver injury that resulted in ineffective deamination by the liver and consequently reduced conversion of amino acid to ammonia, and less conversion of ammonia to urea, the form which is less toxic to the body and majorly excreted through the kidney (24). Also observed was high value of the creatinine levels in all the blood samples collected after the first week of the extract administration for the treated animals (table 1). The test group which got the higher dosage of 1000 mg/kg body weight were found to have an excessive increase in the amount of creatinine level. This finding of elevated creatinine level is in keeping with that previously reported by Afolabi et al who also observed significant increase in creatinine levels at doses higher than 400 mg/kg body weight (10). Also, after two weeks of extract administration (table 2), creatinine levels were found to remain elevated in the test group but values observed were not as high as that recorded after the first week, since creatinine is usually cleared by the kidney and a build-up can indicate some form of renal damage. Unlike urea, the creatinine level was not reverted to the normal value following the termination of extract administration (table 3). Although creatinine is commonly measured as the index of glomerular function, it is not a sensitive index for diagnosis of early renal failure because it may remain within the normal limits even when glomerular filtration rates reduces by half its normal value.

Toxicological examination for other blood and tissue enzyme activity

The result of analysis of other biochemical parameters including alanine transaminase (ALT), aspartate transaminase (AST), alkaline phosphatase (ALP), serum albumin and bilirubin have been provided in tables 1, 2 and 3. Their analysis aided understanding and prognosticating the effect of intake of the ethanolic extract. Enzyme activities of ALT, AST and ALP function as quantitative markers for investigating abnormalies in the organs, especially the liver functions.

After a week of extract administration, values obtained for most of the biochemical parameters were within normal range reported by He et al, for both control and treated animals (26). Similar analysis after two weeks of extract administration revealed significant reduction in levels of bilirubin and reduction in ALT for the treated group at both 500 mg/kg and 1000 mg/kg body weight (table 2). Serum aminotransferase levels (AST and ALT) are two of the most useful measures of possible hepatic cell injury; ALT being more specific is predominantly present in the liver. When the ratio of AST to ALT is greater than 2, the cause of liver injury is likely due to an acute alcoholic hepatitis. As seen in this study for the blood enzyme assay, after two weeks of extract administration at both 500 and 1000 mg/kg body weight, the AST to ALT ratio was found to be much greater than 2 (table 2). This finding suggests possible alcohol toxicity from consumption of the extract. Nkono et al reported mild hepatic and real tissue injury and no significant difference in levels of ALT following administration of aqueous extract from alstonia boonei stem bark (27). The highest dose of extract used for their study was 1000 mg/kg body weight of alstonia boonei stem bark. A similar study has reported no sign of toxicity at doses higher than 200 mg/kg body weight but less than 5000 mg/kg body weight of aqueous extract of alstonia boonei (14). Therefore, the findings from this study suggests that the addition of alcohol increases the toxicity profile of this medicinal extract, hence making it unsafe for consumption as a treatment alternative especially over a prolonged period of time, which is a practise frequently observed in many African countries. The observed low bilirubin levels is asymptomatic and not a cause for concern; as there are no documented health conditions associated with low bilirubin levels, though the consumption of certain substance has been found to reduce bilirubin levels in blood as bilirubin acts as an anti-oxidant that protects tissue against damage from substances that increase cellular destruction.

Table 1: Results obtained from biochemical assay after 7 days[a].

Assay	Control	Low dose	High dose
Albumin (g/L)	35.41 ± 10.04	42.87 ± 6.09	37.35 ± 1.38
ALP (U/L)	214.26 ± 3.39	293.48 ± 90.16	230.74 ± 40.02
ALT (U/L)	52.00 ± 0.00	110.50 ± 43.50	47.50 ± 4.5
AST (U/L)	111.00 ± 7.00	232.00 ± 91.00	82.50 ± 6.5
Bilirubin (μmol/L)	77.80 ± 1.29	119.35 ± 12.67	78.60 ± 3.67
Creatinine (μmol/L)	58.71 ± 8.58	183.00 ± 137.25	345.03 ± 66.72
Urea (mmol/L)	23.66 ± 0.01	22.03 ± 2.29	22.16 ± 0.96

[a]Control specimen was fed distilled water. Low doses signify 500 mg/kg body weight and high dose is 1000 mg/kg body weight. Each value is a mean $\pm$ standard error of duplicates.

Table 2: Results obtained from biochemical assay after 14 days[a].

Assay	Control	Low dose	High dose
Albumin (g/L)	29.82 ± 0.80	32.41 ± 3.39	32.87 ± 1.15
ALP (U/L)	38.64 ± 4.14	121.44 ± 19.52	48.30 ± 23.46
ALT (U/L)	64.50 ± 2.50	8.00 ± 0.00	12.50 ± 4.50
AST (U/L)	89.00 ± 0.00	256.00 ± 9.00	153.50 ± 9.50
Bilirubin (μmol/L)	74.36 ± 0.87	30.58 ± 0.96	21.92 ± 0.40
Creatinine (μmol/L)	47.28 ± 0.89	169.14 ± 19.41	145.13 ± 69.32
Urea (mmol/L)	28.03 ± 1.15	1.67 ± 0.73	2.11 ± 0.42

[a]Control specimen was fed distilled water. Low doses signify 500 mg/kg body weight and high dose is 1000 mg/kg body weight. Each value is a mean ± standard error of duplicates.

Table 3: Results obtained from biochemical assay after 21 days[a].

Assay	Control	Low dose	High dose
Albumin (g/L)	25.31 ± 0.20	35.39 ± 0.68	40.30 ± 1.41
ALP (U/L)	129.03 ± 0.69	376.08 ± 24.84	463.68 ± 80.04
ALT (U/L)	91.00 ± 3.00	100.00 ± 4.00	102.00 ± 2.00
AST (U/L)	139.50 ± 10.50	123.00 ± 0.00	132.00 ± 9.00
Bilirubin (µmol/L)	50.68 ± 0.99	45.68 ± 2.44	47.08 ± 3.16
Creatinine (µmol/L)	56.57 ± 4.36	413.14 ± 152.50	329.96 ± 3.87
Urea (mmol/L)	8.23 ± 0.47	8.51 ± 0.11	9.01 ± 0.27

[a]Control specimen was fed distilled water. Low doses signify 500 mg/kg body weight and high dose is 1000 mg/kg body weight. Each value is a mean ± standard error of duplicates.

DISCUSSION

The major objective of this study was to investigate the toxic effects of ethanolic extract (EE) of the alstonia boonei stem bark on renal function based on the dosage and prolonged period of administration. The benefits of the EE are evident in the rich phytochemicals it contains. The number of phytochemicals found in this study agrees with that previously reported (14) The wide range of these phytochemicals makes the EE suitable for the treatment of several diseases. However, apart from the lack of information about the side effects of this plant EE, and despite its wide-spread use in traditional practice of medicine, there is inadequate information on the underlying biochemical mechanism responsible for some of the observable and reported properties of the plant's stem bark.

Therefore, we have examined the biochemical assay following the extract administration to albino mice. Our findings showed no significant damage to the vital organs or alteration of the biomarkers for the first week when the low dose (500 mg/kg body weight) was administered. This finding is consistent with that previously reported by Nkono et al who found mild hepatic and renal tissue injury and no significant difference in levels of ALT following administration of aqueous extract of alstonia boonei stem bark (27). The highest dose of extract used for their study was 1000 mg/kg body weight of alstonia boonei stem bark. The major difference between this study and theirs is the presence of ethanol in our extract. Also, a similar study reported no sign of toxicity following consumption of the EE at doses higher than 200 mg/kg body weight but less than 5000 mg/kg body (14). A reduction in the levels of urea and increase in creatinine was observed when the extract was administered for two weeks, both in high and low doses. This finding of elevated creatinine level agrees with that previously reported by Afolabi et al (14) who also observed significant increase in creatinine levels at doses of EE higher than 400 mg/kg body weight. Also observed were some changes in levels of other blood and tissue enzyme activity. After the first week of ethanolic extract administration, levels of ALT, AST, ALP, serum albumin and bilirubin were essentially normal (26,29,30). After the second week, the biochemical assay revealed a significant reduction in levels of ALT and bilirubin. A striking observation in this study is that the ratio of AST to ALT was greater than 2 after two weeks of extract administration at both 500 and 1000 mg/kg body weight, which suggests a possible alcohol toxicity from consumption of the ethanolic extract.

CONCLUSION

The toxicological effects of alstonia boonei stem bark based on duration and dosage of administration have been investigated. Ethanolic extract was found to be toxic at therapeutic doses when consumed over a prolonged period (especially for 2 weeks). A low dose of the extract was found to be a somewhat safe alternative for the treatment and prevention of common illnesses over a short period time as no significant damage to vital organs was observed when 500 mg/kg body weight was administered for one week. However, their prolonged use is unsafe and not recommended. The unwanted effect of the alcohol in the extract makes it more toxic to the safety profile obtained when a similar dose of the aqueous extract is administered. The risk of taking the ethanolic extract of alstonia boonei stem bark for a prolonged period appeared to far outweigh the benefit derived from its consumption.

REFERENCES

1. Sachs J, Malaney P. The economic and social burden of malaria. Nature 2002; 415:680-685.

2. Ginsburg H, Deharo E. A call for using natural compounds in the development of new antimalarial treatments–an introduction. Malar J 2011; 10:1

3. Odugbemi TO, Akinsulire OR. Medicinal plants useful for malaria therapy in Okeigbo, Ondo State, Southwest Nigeria. Afr J Trad CAM 2007; 4: 191-198.

4. Taylor WR, White NJ. Antimalarial drug toxicity: a review. Drug Saf 2004;27:25-61.

5. Cruz LR, Spangenberg T, Lacerda MVG, Wells TNC. Malaria in South America: a drug discovery perspective. Malar J 2013; 12:168.

6. Rezaei N, Alinia P, Aghabiklooei A, Izadi Sh. Blood Lead Level in Opium Abuse; Which Is More Dangerous? Opium Smoking or Opium Ingestion? Asia Pac J Med Toxicol 2019;8(4):124-9.

7. Bello I S, Oduola T, Adeosun OG, Omisore NOA., Raheem GO, Ademosun A. Evaluation of Antimalarial Activity of Various Fractions of Morinda lucida Leaf Extract and Alstonia boonei Stem Bark. Global J Pharmacol 2009;3: 163-165.

8. Chime SA, Ugwuoke EC, Onyishi VI, Brown SA., Onunkwo GC. Formulation and Evaluation of Alstonia boonei Stem Bark Powder Tablets. Indian J Pharm Sci 2013; 75:226-230

9. Yenesew A, Induli M, Derese S, Midiwo JO, Heydenreich M, Peter MG, Akala H, Wangui J, Liyala P, Waters NC. Anti-plasmodial flavonoids from the stem bark of Erythrina abyssinica. Phytochemistry 2004; 65:3029–3032.

10. Afolabi BA, Obafemi TO, Akinola TT, Adeyemi AS, Afolabi OB. Effect of ethanolic extract of alstonia boonei leaves on serum electrolyte levels in wistar albino rats. Pharmacologyonline 2015; 3:85-90.

11. Ahsan MR, Islam KM, Bulbul IJ, Musaddik MA, Haque E. Hepatoprotective activity of methanol extract of some medicinal plants against carbon tetrachloride-induced hepatotoxicity in rats. Eur J Sci Res 2009; 37: 302 – 310.

12. Wichmann O, Muhlen M, Grub H, Mockenhaupt FP, Suttorp N, Jelinek T. Malarone treatment failure not associated with previously described mutations in the cytochrome b gene. Malaria J 2004; 3: 1-3.

13. Olajide OA, Awe SO, Makinde JM, Ekhelar AI, Olusola A, Morebise O, Okpako DT. Studies on the anti-inflammatory, antipyretic and analgesic properties of Alstonia boonei stem bark. J Ethnopharmacol 2000; 71: 179–186.

14. Akinmoladun AC, Ibukun EO, Afor E, Akinrinlola B L, Onibon TR, Akinboboye O, Obuotor EM Farombi EO. Chemical constituents and antioxidant activity of Alstonia boonei. Afr J Biotech 2007; 6: 1197 – 1201.

15. Pinnell AE, Northam B E. New automated dye-binding method for serum albumin determination with bromcresol purple. Clin Chem 1978; 24: 80-86

16. Reitman S, Frankel S. A colorimetric method for the determination of serum glutamic oxalacetic and glutamic pyruvic transaminases. Am J Clin Pathol 1957; 28:56-63

17. Kume T, Saglam B, Ergon C, Sisman AR. Evaluation and comparison of Abbott Jaffe and enzymatic creatinine methods: Could the old method meet the new requirements? J Clin Lab Anal 2018; 32:e22168.

18. Sanni MS, Van-Etten RL. An essential carboxylic acid group in human prostrate acid phosphatase. Biochim. Biophys Acta 1978; 568:370-376.

19. Gambino SR. Bilirubin (modified Jendrassik and Grof)-provisional. Stand. Methods Clin Chem 1965; 5: 55-64.

20. Gordon SA, Fleck A, Bell J. Optimal conditions for the estimation of ammonium by the Berthelot reaction. Ann Clin Biochem 1978; 15: 270-275.

21. Nnyigide OS, Oh Y, Song H, Park E, Choi S, Hyun K. Effect of urea on heat-induced gelation of bovine serum albumin (BSA) studied by rheology and small angle neutron scattering (SANS). Korea-Aust Rheol J 2017; 29: 101-113.

22. Bray GA, Preston AS. Effect of urea on urine concentration in the rat. J Clin Invest 1961; 40: 1952–1960.

23. Roberts DM, Gallapatthy G, Dunuwille A, Chan BS. Pharmacological treatment of cardiac glycoside poisoning. Br J Clin Pharmacol 2015; 81: 488-95.

24. Weiner ID, Mitch WE, Sands JM. Urea and Ammonia Metabolism and the Control of Renal Nitrogen Excretion. Clin J Am Soc Nephrol 2015; 10: 1444–1458.

25. Nnyigide OS, Lee SG, Hyun K. Exploring the differences and similarities between urea and thermally driven denaturation of bovine serum albumin: intermolecular forces and solvation preferences. J Mol Model 2018; 24: 75

26. He Q, Su G, Liu K, Zhang F, Jiang Y, Gao J et al. Sex-specific reference intervals of hematologic and biochemical analytes in Sprague-Dawley rats using the nonparametric rank percentile method. PLoS ONE 2017; 12: e0189837.

27. Nkono BLNY, Sokeng SD, Desire DDP, Kamtchouing LFP. Subchronic toxicity of aqueous extract of Alstonia boonei de wild. (apocynaceae) stem bark in normal rats. Int. J Pharmaco Toxicol 2015; 3: 5-10.

28. Nnyigide OS, Hyun K. Effect of urea and temperature on the molecular dynamics of bovine serum albumin in heavy and light water. J Chem Technol Metall 2016; 51, 147.

29. Winter RW, Kelly JX, Smikstein MJ, Dodean R, Bagby GC. Evaluation and lead optimization of antimalarial acridones. Exp Parasitol 2006; 114: 47-56.

30. Mcdonough M. Alcohol Use Disorders: Implications for the Clinical Toxicologist. Asia Pac J Med Toxicol 2015;4:13-24